DID YOU KNOW?
ANIMAL ADAPTATIONS

20 THINGS YOU DIDN'T KNOW ABOUT

INSECT ADAPTATIONS

SLOANE HUGHES

PowerKiDS press
New York

Published in 2023 by The Rosen Publishing Group, Inc.
29 East 21st Street, New York, NY 10010

Portions of this work were originally authored by Kristen Rajczak and published as *20 Fun Facts About Insect Adaptations.* All new material in this edition was authored by Sloane Hughes.

Editor: Amanda Vink
Book Design: Tanya Dellaccio

Photo Credits: Cover (speech bubble) Gohsantosa/Shutterstock.com; Cover (logo art) NazArt/Shutterstock.com; Cover (Butterfly) Paul Sparks/Shutterstock.com; p. 5 (praying mantis) Lightspring/Shutterstock.com; p. 5 (shield bug) YapAhock/Shutterstock.com; p. 5 (bee) Viesturs Ozolins/Shutterstock.com; p. 5 (ant) Witsawat.S/Shutterstock.com; p. 6 Digital Media Pro/Shutterstock.com; p. 7 Good Shop Background/Shutterstock.com; p. 8 Norjipin Saidi/Shutterstock.com; p. 9 Tacio Philip Sansonovski/Shutterstock.com; p. 10 Daisuke Nishioka JP/Shutterstock.com; p. 11 Steve Bower/Shutterstock.com; p. 12 YoONSpY/Shutterstock.com; p. 13 Natalia Kuzmina/Shutterstock.com; p. 14 Sean McVey/Shutterstock.com; p. 15 Loc'd Lense/Shutterstock.com; p. 16 StudioSmart/Shutterstock.com; p. 17 (butterfly) ilikestudio/Shutterstock.com; p. 17 (life cycle) Nicolas Primola/Shutterstock.com; p. 18 Johan Swanepoel/Shutterstock.com; p. 19 Jan Mastnik/Shutterstock.com; p. 20 keshavadav1111/Shutterstock.com; p. 21 fish1715/Shutterstock.com; p. 22 Bildagentur Zoonar GmbH/Shutterstock.com; p. 23 schankz/Shutterstock.com; p. 24 skynetphoto/Shutterstock.com; p. 25 Paul Reeves Photography/Shutterstock.com; p. 26 yod 67/Shutterstock.com; p. 27 junrong/Shutterstock.com; p. 29 apiguide/Shutterstock.com.

Library of Congress Cataloging-in-Publication Data
Names: Hughes, Sloane, author.
Title: 20 things you didn't know about insect adaptations / Sloane Hughes.
Other titles: Twenty things you didn't know about insect adaptations
Description: New York : PowerKids Press, [2023] | Series: Did you know? animal adaptations | Includes index.
Identifiers: LCCN 2021055257 (print) | LCCN 2021055258 (ebook) | ISBN 9781538386798 (library binding) | ISBN 9781538386774 (paperback) | ISBN 9781538386781 | ISBN 9781538386804 (ebook)
Subjects: LCSH: Insects–Adaptation–Juvenile literature. | Adaptation (Biology)–Juvenile literature.
Classification: LCC QL467.2 .H845 2023 (print) | LCC QL467.2 (ebook) | DDC 595.7-dc23/eng/20211118
LC record available at https://lccn.loc.gov/2021055257
LC ebook record available at https://lccn.loc.gov/2021055258

Manufactured in the United States of America

CPSIA Compliance Information: Batch #CSPK23. For Further Information contact Rosen Publishing, New York, New York at 1-800-237-9932.

CONTENTS

INSECTS EVERYWHERE!

Insects are very good at surviving. They make up about two-thirds of all the creatures **identified** so far by humans on the planet. There are roughly 1 million species, or kinds, of insects. They come in as many shapes, sizes, and colors as you can imagine!

Insects are found on every continent. They have body and behavioral **adaptations** that allow them to survive in different **habitats**. These adaptations are sometimes weird, and they're always amazing. Let's learn more about insects!

All insects have six legs and three basic body parts.

INSECT BODY PARTS

DID YOU KNOW?

Insect mouthparts are very **unique**.

The shape of an insect's mouth depends on what it likes to eat. Crickets and other insects that eat plants or other insects have mouthparts for chewing. These are called mandibles. Butterflies, on the other hand, have mouths shaped like straws, which they use to drink **nectar**.

Like many animals on Earth, different species of insects have different adaptations depending on where and how they survive.

DID YOU KNOW?

Some ants use sharp mandibles to cut leaves.

You may wonder what these ants do with this perfectly cut greenery! The ants carry the cut leaves to an underground nest, where they grow gardens of **fungi**. That's what the larvae, or young insects, eat.

DID YOU KNOW?

Insects change their armor as they grow larger.

Insects have hard outer shells called exoskeletons. Exoskeletons are like armor for insects! As an insect grows, its exoskeleton becomes too small. When that happens, an insect can shed its outer skin in order to grow a new one.

When an insect sheds its exoskeleton, that's called molting. It may happen many times throughout an insect's life.

DID YOU KNOW?

Some insects have big eyes and small antennae, or feelers. Some insects have small eyes and large antennae.

Insect eye size and antennae length depend on the brightness of the insect's home. If an insect lives in a dark place, it doesn't need to see well. But it needs long antennae to feel around! On the other hand, if an insect lives in a bright place, its eyes need to be big and its antennae don't need to be as long.

DID YOU KNOW?

Dragonflies have wraparound vision.

Dragonflies have two compound eyes, each with several thousand parts. This creates an image much different from what we see. Creatures with compound eyes are great at noticing differences in light, which helps them **navigate** while flying.

INSECT TRICKS

DID YOU KNOW?

The orchid mantis has legs that look like flower petals.

The orchid mantis has very effective **camouflage**. That's so predators and prey alike can't see it. This mantis looks like a flower called an orchid, which allows it to hide in plain sight.

Can you find the orchid mantis?

CAMOUFLAGE	THE INSECT BLENDS INTO ITS SURROUNDINGS AND TRIES TO HIDE.	WALKING STICK, ORCHID MANTIS, KATYDID, DEAD LEAF MANTIS, SAND GRASSHOPPER, PEPPERED MOTH, THORN BUG
MIMICRY	THE INSECT LOOKS LIKE SOMETHING ELSE TO SCARE OFF PREDATORS.	VICEROY BUTTERFLY, ROBBER FLY, LADYBIRD MIMIC BEETLE, WASP BEETLE, BEE BEETLE, AMERICAN HORNET MOTH

DID YOU KNOW?

Stick insects, also called walking sticks, can drop their legs off to escape!

Walking sticks also blend in with their surroundings. They even sway back and forth with the wind in order to copy twigs. But if they're caught, they can lose a leg to get away and grow it back later.

IN THE COLD

DID YOU KNOW?

Some insects stop growing, eating, and moving in the cold.

When conditions outside are really icy, some insects shut down for a while. This is called diapause, and it often happens in the winter for insects that live in places with four seasons.

Yellow jacket queens often go into diapause during winter. Upon waking up in spring, the queen leaves to start a new nest.

These species are able to survive because their bodies make cryoprotectants, **substances** that protect against damage caused by very cold temperatures. Some woolly bear caterpillars can withstand temperatures as low as -70°Fahrenheit (-57°C).

DID YOU KNOW?

Some insects are built for extremely low temperatures.

DID YOU KNOW?

Honeybees work together to stay warm.

When temperatures drop, honeybees begin to gather together in their hive. There are some bees whose job is to shiver their muscles quickly to **generate** heat. The queen and the eggs stay in the center of the group, safe and warm.

Honeybees rely on their adaptations to survive. Their coloring works as camouflage. They have a stinger to fight off predators.

DID YOU KNOW?

Monarch butterflies that become adults in late summer go through a type of diapause over winter. Because of this, they live longer than other monarchs.

In the fall, monarch butterflies migrate 3,000 miles (4,828 km) to Mexico, where it's warmer. In the spring, they're able to **reproduce** and their offspring will continue the rest of the way north for the summer.

WACKY HABITATS

DID YOU KNOW?

Dung beetles live on, and in, animal waste.

Dung beetles spend their life eating other animals' poop, or dung. Special spikes on their back legs help them roll dung into little balls. Their strong front legs are perfect for digging. Adult dung beetles lay their eggs in dung balls.

DUNG BEETLE LARVA

In a way, dung beetles are the world's strongest animal. They can roll dung balls that are more than 1,000 times their own weight!

DID YOU KNOW?

Young dragonflies and damselflies have gills so they can live underwater!

When baby dragonflies and damselflies hatch, they're in or very near the water. They use gills to breathe in water. They can even shoot a jet of water out of their gills in order to speed away from danger. As they get older, dragonflies and damselflies live out of the water.

Young dragonflies and damselflies molt when they leave the water. Within a few hours, their bodies harden and they are ready to fly.

DID YOU KNOW?

Giant water bugs have long claws they use to snatch prey.

Insects that spend their lives in water need special adaptations. Giant water bugs have strong front legs with sharp claws for catching underwater prey. They also have back legs that are great for swimming.

OTHER ADAPTATIONS

DID YOU KNOW?

The mole cricket has shovel-shaped legs that help it dig through soil.

Mole crickets make tunnels using their toes, which are called dactyls. Then they use their heavily armored bodies to shape and pack the soil.

When mole crickets dig through dirt, it's called tunneling.

DID YOU KNOW?

Fleas are so tiny that other animals can barely feel them feeding.

Fleas have a flat body. They don't have wings, and they have short antennae. Fleas also have really tiny mouthparts. These adaptations allow the flea to hide and feed on another animal's body.

DID YOU KNOW?

Cockroaches have six legs, but they run fastest on their back legs alone!

Cockroaches have long, thin legs. Each of their six legs has three knee. Cockroaches can run up to 50 body lengths in one second! Once they get up to speed, they rise on their hind legs and run like people do.

That's about 30 inches (76.2 cm)! Grasshoppers have long and strong back legs made for jumping. Their hind legs are built like springs.

DID YOU KNOW?

Honeybees die after they sting, but they do it to protect the hive.

Honeybee stingers have hooks on them. The stinger gets stuck in an animal's skin, and it often rips away the back end of the bee. The honeybee might fly away, but it won't get far. Honeybees don't sting unless they feel **threatened**. Usually, it's to defend their nest.

Stink bugs arrived in the United States in the 1990s.

When stink bugs are afraid, they spray stinky liquid. This warns predators to stay away! Some stink bugs can even shoot the liquid several inches away.

DID YOU KNOW?

Stink bugs really do stink!

SMALL BUT MIGHTY!

Insects may be small, but they are mighty! They are the most plentiful type of animal on the planet. Insects have found a way to live all over the world, thanks to their amazing adaptations!

While insects share many **traits**, they are different in so many ways. They come in different sizes, shapes, and colors. Next time you're outside, see how many insects you can spot. What makes them different from one another? And what adaptations help them survive?

Scientists estimate that there are at least 2 million insects per human being on Earth.

GLOSSARY

adaptation: A change in a living thing that helps it live better in its habitat.

camouflage: Colors or shapes on animals that allow them to blend in with their surroundings.

fungus: A living thing that's like a plant but that doesn't have leaves, flowers, or green color or make its own food. The plural form is fungi.

generate: To produce something or cause something to be produced.

habitat: The natural home for plants, animals, and other living things.

identify: To find out or show the identity of.

navigate: To find one's way.

nectar: A sweet liquid produced by plants and used by bees in making honey.

reproduce: To produce another living thing of the same kind.

substance: A material of a particular kind.

threatened: Likely to be harmed.

trait: A quality that makes one person, animal, or thing different from another.

unique: Special or different from anything else.

FOR MORE INFORMATION

BOOKS

Messner, Kate. *Insect Superpowers*. San Francisco, CA: Chronicle Books, 2018.

Riehecky, Janet. *Insects*. Oxford, UK: Raintree, 2018.

WEBSITES

DK Find Out: Insects
www.dkfindout.com/us/animals-and-nature/insects/
With this interactive site about all kinds of insects, you'll be sure to find some cool bug facts!

Ducksters: Butterflies
www.ducksters.com/animals/butterfly.php
Want to learn more about butterflies? This site will give you plenty of information about monarchs and much more.

Publisher's note to educators and parents: Our editors have carefully reviewed these websites to ensure that they are suitable for students. Many websites change frequently, however, and we cannot guarantee that a site's future contents will continue to meet our high standards of quality and educational value. Be advised that students should be closely supervised whenever they access the internet.

INDEX

TITLES IN THIS SERIES

20 THINGS YOU DIDN'T KNOW ABOUT
AMPHIBIAN ADAPTATIONS

20 THINGS YOU DIDN'T KNOW ABOUT
BIRD ADAPTATIONS

20 THINGS YOU DIDN'T KNOW ABOUT
INSECT ADAPTATIONS

20 THINGS YOU DIDN'T KNOW ABOUT
MAMMAL ADAPTATIONS

20 THINGS YOU DIDN'T KNOW ABOUT
REPTILE ADAPTATIONS

20 THINGS YOU DIDN'T KNOW ABOUT
SEA ANIMAL ADAPTATIONS

PowerKiDS press

ISBN: 9781538386774
6-pack ISBN: 9781538386781
9 781538 386774